Three Pz.Kpfw.B2 (Flamm) outside Le Sénat. They are seen here facing towards the exit on Rue de Médicis where the Panther on pages 6-7 was knocked out. The flame gun replaced the 7·5cm Kw.K.35(f) in the front hull and the flame operator was provided with the standard German 'Fahrersehklappe 50' for vision. Note the cap still on the 4·7cm turreted weapon and original cupola.
L.Archer

A Pz.Kpfw.17R 730 (f), with Luftwaffe licence plate, stands alone in Le Sénat, as photographed a US Army Signal Corps photographer on 27 August. Note the civilian truck in the left of the picture covered with camouflage net. On the right is an Italian SPA CL39.

NARA

The same Pz.Kpfw.17R 730 (f), now in a new location and stripped of equipment such as weapon, jack, tools, etc. The tank is painted in a single colour - probably RAL 7021 'Dunkelgrau.' These little tanks, that traversed WWI battlefields, were replaced by the Renault R35 which, in German service, had trouble negotiating similar terrain, and were mainly used in rear areas.

L.Archer

\ Selbstfahrlafette mit Panzerung (Sd. ‹fz.11/1) für 2cm Flak 38 inside the .ourtyard. The armoured cab provided :over for the driver, gun commander ınd four 'Kanoniere' who serviced :he gun. The space between the gun ›latform and the top of the frame was ıot armoured. The rear of the platform was taken up by munitions lockers :opped by a flat, rectangular box for :he spare 2cm Flak barrels. The vehicle was used by le.Fla.Abt.98 (gep, SF) of III. ⁼lakkorps. The Pz.Sp.Wg.Panhard 178 ᵓ204 (f) belonged to 14.Aufklärungs ‹ompanie. **3x NARA, 1x L.Archer**

This Panther, seen in front of the Théâtre de l'Odéon on Rue de Vaugirard , was one of only two Panthers to put up a fight in Paris. Hemmed in by Shermans of 2nd Co. 501e RCC, it moved back and forth in the smoke of shell fire and was struck four times by 'La Moskowa,' a 105mm armed Sherman, after which the rattled crew bailed out. One of the impacts blew off a chunk of armour next to the driver, visible on the opposite page. **2x L.Archer**

Both sides of a Sfl 7·5cm Pak 40 auf Fahrgestell 38H 735 (f) that was built by Baukommando Becker and knocked out by the French Résistance with a 'molotov cocktail' on Rue de Courcelles. The vehicle had the dubious honour of being the first tank destroyed in Paris by the French Résistance.
2x L.Archer

ESS
PRESSING
Gaby

Another casualty of a French Résistance molotov cocktail, courtesy of Michel Aubry, who later became the curator of the famed Musée Blindés in Saumur. This Panther Ausf.G, assembled by Daimler-Benz in June 1944, was travelling past the Hôtel de Ville when it was 'cocktailed.' It continued to Quai Henri IV where the crew, unable to extinguish the fire, bailed out. Like many Panthers and Tigers shown in *Panzerwrecks*, the torsion bar suspension has sagged as a result of the fire. The 'Fliegerbeschussgerät' ring on the commander's cupola has been removed.

2x L.Archer

As if storming the ramparts, these Résistance fighters have ascended a Matford LKW to secure their flag in the upturned barrel of a 2cm Flakvierling. Considered a heavy truck in the 4.5 t range, the Matford F917-WS should have been able to shoulder the load easily, but being a 4x2, it was best on roads.

L.Archer

A Pz.Kpfw.35S 739 (f) captured by the F.F.I. (Forces Francaises de l'Interieur) at the Somua factory in Saint-Ouen, a Paris suburb. Without ammunition, the tank had only a symbolic role during the uprising and appears to have been turned into a canvas or 'yearbook,' with several numbers ('34,' '29,' '24,' etc.), tricolours, multiple 'Croix de Lorraine' and the name 'MOUNET' among the inscriptions. 'Mounet' may refer to Paul Mounet, a famous actor who appeared at the Odéon in Paris who died in 1922. Our resistance fighters brandish an assortment of weapons, including the feared Gallic cigarette. Note the car in the left background has also received a number. **L.Archer**

TOUT POUR LA T.S.F.
BEAUSOLEIL
AU PARADIS DES
CAFÉ

\ turretless Pz.Kpfw.B2 als Schulfahrzeug (driver training vehicle) outside Église Saint-Paul on Rue Saint-Antoine. Along with the removal of the turret, the Germans removed the 7·5cm gun from the hull, covering over the hole with steel plate, on which a Parisian has written "A vendre Prix a Debattre" (Sale price for discussion). **1x L.Archer, 1x NARA**

The same vehicle photographed from the steps of Église Saint-Paul, with a clutch of Parisians straining to get a look into the side hatch of the tank. Its size, even without turret, is evident here against the height of the people. An intrepid pair are checking out the driver's position and the engine bay. **L.Archer**

A 3·7cm Flak 37 auf Selbstfahrlafette (Sd.Kfz.7/2) knocked out by the 7th Co, IIRMT and later shot up by M4s of the 12e Régiment de Chasseurs d'Afrique in Jouy-en-Josas. Note how the roof of the cab has been peeled back. Armour plate thickness was just 8mm.

N.Conreur

'wo more views of the Jouy-en-Josas Flakwagen. The deathblow, from the M4, has ripped ›ff the roof from the armoured cab. (Note that this roof had two hatches built into it.) A detail not often seen is that of the punched steel plate that makes up the drop-down sides for the 'fighting compartment.' We can see that the inner surface protrudes to allow more grip for the five 'Kanoniere' serving the 3·7cm gun. In the background we can see the wide expanse of sky it overlooked. **2x N.Conreur**

Almost unrecognizable, this armoured Sd.Kfz.7/2 was knocked out at the entrance of Clamart by navy infantry (*Infanterie de Marine)* of the 7th Co, IIRMT (*Regiment de Marche du Tchad*). It was finished off by M4s of 12e RCA (*Regiment de Chasseurs d'Afrique*). Date 24 August 1944.

N.Conreur

.ess than a mile away from the Flakwagen on pages 17-19 was this Jagdpanzer IV ›elonging to SS-Pz.Jg.Abt.12. It was knocked out by 'Cyclone', an M10 Tank Destroyer of he 3rd Platoon, 4th Squadron of the RBFM (*Regiment Blinde de Fusiliers-Marin*) on the ›vening of 24 August. The vehicle had been travelling from Versailles to Clamart along the N186 when hit. Badly scorched by fire, it was later pushed off the road, sustaining further damage to its left side. Unlike the standard Pz.Kpfw.IV, the Jagdpanzer IV had both left and right side track guard supports.

N.Conreur

The nose armour on this example was 60mm thick, it was later increased to 80mm starting with the 301st vehicle. Here the tapered brackets for the 'Schürzen' and threaded end to the gun barrel are seen to good advantage. The conical cover for the right side MG port is neither fully opened nor fully closed. **Opposite page:** It seems it was quite an attraction, roller coaster tracks and all.

2x N.Conreur, 1x W.Auerbach, 1x L.Archer

Above: A Panther Ausf.A knocked out in the Jardin des Tuileries by an M4A3 named 'Francheville' of 501e RCC. **Right and far right:** Another dead Panther in the Jardin des Tuileries; this one an Ausf.G knocked out by an M4A2 named 'Surcouf' of 501e RCC.
3x L.Archer

Right and bottom right: Jardin des Tuileries. Pz.Kpfw.35R 731 (f) No.'52' at the junction of Rue de Rivoli and Rue d'Alger. It was knocked out by two bazooka rockets from soldiers of 3rd Co IRMT. One struck the left engine hatch and blew it off, the second struck just above the right track.

2x L.Archer

Above: Boys will be boys, and the urge to spin the turret of the Pz.Kpfw.35R 731 (f) must have proved irresistible. One boy peers down the split hatch on top while another goes through the drop down hatch in the rear. Small details, like the handles on the engine hatches, set this vehicle apart from the one depicted on pages 26-27. This photograph was taken sometime later after the tank had been moved further down Rue d'Alger, and in fact stands outside property number 6, with its ornate entrance.

L.Archer

RUE DU
MONTTHABOR
51

51
51

Page 26-28: Again in the vicinity of Jardin des Tuileries. Pz.Kpfw.35R 731 (f) No. '51' left on the corner of Rue du Mont Thabor and Rue de Castiglione. 'Mont Thabor' was a battle fought by Napoleon's army in Syria almost a century and a half earlier, and one starts to grasp the sense of history built into Paris' DNA. In the distance stands the Colonne Vendôme made of guns captured by Napoleon's army. **Left**: A rear view of '51'. This photo, unlike others in this book, has not been 'cleaned', to show the 'found' condition of some of our photos. Captured Renault R35 were fielded in three versions: unmodified; fitted with radio and Antennafuss for the 1·4 metre antenna on the left side and with the top of the commander's cupola cut off and two piece hatch installed; and with the radio, Antennafuss and cupola completely removed and a large two part hatch installed directly to the turret roof.

4x L.Archer

Above right and opposite: Another scene on Rue de Castiglione facing towards the Pz.Kpfw.35Rs and the Jardin des Tuileries and it's a picture of complete immolation. Parisian cyclists *regardent la voiture* while other people mill about the mittlerer Schützenpanzerwagen converted to a Funkwagen, both destroyed by tanks of the 1st Section, 3rd Co. 501e RCC. The photo above shows entry and exit holes on the right hand side.

1x W.Auerbach, 1x L.Archer

Left: City folk scrounge through the burnt out interior of the converted m.SPW from page 29 as even small amenities disappeared during the occupation. The 'Sternantenne' had already replaced the frame antenna on the final production of the Sd.Kfz.251/3 Ausf.C and radio sets and racks are absent. **Opposite**: The stanchions that supported the large antennas in these vehicles were made of wood, to act as insulators, and here they have burned away. Note the scorched paint and larger than usual 'Balkenkreuz'. The tray like objects on the ground may have held accessories and spares for the radios. **2x L.Archer**

Above: A *gendarme* feels the heft of an MG42 still mounted on a m.SPW (Sd.Kfz.251/1), perhaps not realising the metal hoop in front of him that supported the vehicle's canvas cover was removable. A Signal Corps motion picture cameraman recorded this scene and the ones on pages 32-33. The SPW was emblazoned with large white stars, indicating it was captured prior to rolling into Paris. The two men on page 33 wear berets with badges showing two crossed guns, the headgear of French armoured units in 1939. The man on the left also has the armband typical of the French Resistance.

NARA

Resistance fighters occupy the captured m.SPW (Sd.Kfz.251/1) shown on the previous page. The SPW is now more of a curiosity, as what they are really looking for is a return to normality. A partial number '43' is visible between the stowage boxes. The 'crew' have fitted a French 7·5mm Kpfw MG 311(f) to the rear pintle. **2x NARA**

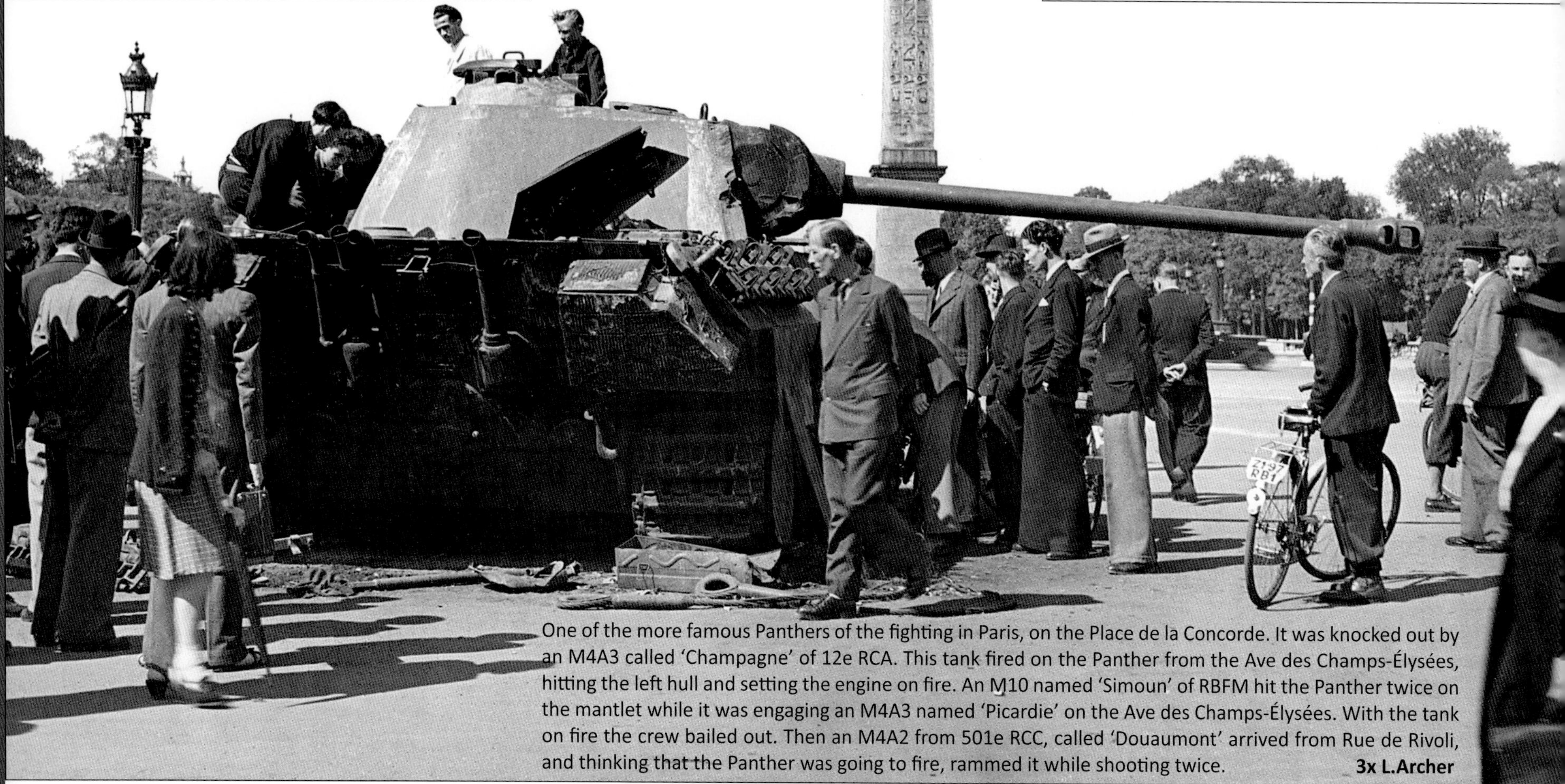

One of the more famous Panthers of the fighting in Paris, on the Place de la Concorde. It was knocked out by an M4A3 called 'Champagne' of 12e RCA. This tank fired on the Panther from the Ave des Champs-Élysées, hitting the left hull and setting the engine on fire. An M10 named 'Simoun' of RBFM hit the Panther twice on the mantlet while it was engaging an M4A3 named 'Picardie' on the Ave des Champs-Élysées. With the tank on fire the crew bailed out. Then an M4A2 from 501e RCC, called 'Douaumont' arrived from Rue de Rivoli, and thinking that the Panther was going to fire, rammed it while shooting twice. **3x L.Archer**

ᵓages 35-36: The scene outside Hôtel Majestic on Avenue Kléber. Two Renault R40s (an ımproved R35 with better main gun and suspension) and a 4·7cm Pak (t) (Sfl.) auf Fgst. ᵓz.Kpfw. 35R (f) have been knocked out by an M10 from RBFM called 'Mistral' firing down the avenue from the Arc de Triomphe. The Parisians on page 36 are helping themselves to German fuel. Another of the diminutive tank hunters was knocked out by a bazooka close by on Rue Pauquet, now renamed Rue Jean Giraudoux. **1x L.Archer, 1x NARA**

The shattered remains of a Sturmgeschütz III Ausf.G lie in Soisy-sous-Montmorency, to the North of Paris. It was destroyed by an M10 named 'Ouragan' of 4th Squadron RBFM on 27 August by a round into the side, detonating the ammunition. This StuG, and two others knocked out nearby, were probably part of Sturmgeschütz-Brigade.341, which was refitting in the Paris suburbs.

L.Archer

In a city known for its beauty, this burnt out Jagdpanzer IV is indeed an eyesore. This vehicle was actually destroyed by its crew on the outskirts of Paris, on Avenue Galliéni, in Épinay-sur-Seine. Note the brackets on the side of the fighting compartment and blown barrel. The left MG port was dropped in March 1944. **L.Archer**

The same vehicle from the other side, facing towards Enghien-les-Bains, a half mile away. The busted barrel clearly identifies it as the same vehicle as on the previous page. The Jagdpanzer IV was about as tall as the average Frenchman, and its subdued shading and non-reflective 'Zimmerit' coating would have made a stationary vehicle tough to spot. A patrol from 2e DB came upon this vehicle on 26 August. Its roof is slightly ajar, probably from the crews' demolition charge. Like the example on pages 21-23, it is from SS-Pz. Jg.Abt.12.

L.Archer

A Jagdpanther photographed by a member of the 82nd Airborne Division, probably with the 508th Parachute Infantry Regiment. It appears to have been pushed off the road onto a stream bank. It has a tiger stripe paint scheme, bolted mantlet, mounts for a rangefinder, and its exhaust pipes seem to go straight up. It is missing the two piece loader's hatch, traversable periscope for the commander, fuel and coolant filler caps, several road wheels and any sign of 'Zimmerit.' The engine hatch is open. Neither the Jeep nor the Jagdpanther offer any identifying clues as to unit or location.

2x B.Gauthier

Two views of a Tiger Ausf.B with broken 'Verladekette' (transport track). A futile attempt had been made to tow this vehicle and get its massive fire power back in action. Part of the engine decking with the cooling air outlet hangs from the rear. 'Zimmerit' was dropped in September, 1944. A front view of this vehicle appeared on page 29 of *Panzerwrecks 1*, and a side view appeared later on page 15 of *Panzerwrecks 5*, and it was identified as being from 3./s.H.Pz.Abt.511. **2x W.Auerbach**

G.I.s pose with Tiger Ausf.B '134' from s.Pz.Abt.506 at Schmallenberg, Germany. For reasons unknown this vehicle broke down on 5 or 6 April 1945. The crew had set off demolition charges, the resulting fire scorching the gun barrel, turret and sagging the torsion bar suspension. The '463' marking painted on the track guard means that the US 463rd Ordnance Evacuation Company would have been responsible for removing the wreck from the battlefield.
J.Haley

An armoured bulldozer from the 237th Engineer Combat Battalion pushes the hulk of a Tiger Ausf.B from s.Pz.Abt.507 off the road at Harste, North of Göttingen, Germany. The tank was one of four knocked out during a counterattack on Harste on 9 April 1945. The tank was assembled by Henschel in March 1945 and features 18 toothed drive sprockets with single link tracks and reinforcing ribs on the front track guards.

NARA

Karlsplatz, Munich. August 1945. M.A.N. in Nuremberg was producing Panthers with a 800mm resilient steel wheel installed on the last road wheel station in March and April 1945, and it is believed that this example was taken directly from the assembly line and brought to this location by American troops for eventual shipment to the Patton Museum at Fort Knox, Kentucky. Note that the track run does not touch the top of the steel wheel since it is 60mm smaller in diameter than the normal road wheels. **Opposite page:** A superlative close-up view of the vehicle and what may be members of its escort to the railhead and beyond.

1x W.Auerbach, 1x B.Gauthier

imperial
FILMTHEATER

Opposite: Some front end damage has occurred to the rail for hanging the 'Schürzen.' It is amazing that such clear photos were taken of both sides of the same vehicle. Spare track links, but none of the outer vehicle equipment such as crowbars and axes, etc., have been fitted to the sides of the rear superstructure. This unique view shows the factory applied camouflage scheme and the loops for attaching foliage on the turret. Note that the MG rail for the commander's cupola was no longer fitted. This Panther carried Fahrgestell Nr. 121455. **This page:** General Patton and his entourage photographed as they pass by a Pz.Kpfw.III Ausf.F (that also went to the Patton Museum) with our Munich Panther in the background.

1x USAHEC, 1x L.Archer

Here are two more photos of the Panzerbefehlswagen III Ausf.D1 that first appeared on pages 88-89 of *Panzerwrecks X*, and they do a nice job of rounding out the picture of this early attempt at producing a command vehicle for the Panzertruppen. The hatches to access the fan drives and engine compartment are all open. Note the stand-off bracket for the smoke candle discharger, jack supporting the rear of the vehicle and the cut out in the rear section of track guard for the convoy light. Two Varta 12 volt batteries were carried and appear here dismounted. Note how the bottom of the rear armour has a slight 'dip' in the centre instead of being straight across.

W.Auerbach

The skinny 'Balkenkreuz' can't hide the weight gain; the command version was over two metric tons heavier than the normal Pz.Kpfw. due to an armour upgrade from 14·5mm to 30mm. This thickening can be seen most readily in the cover of the cooling air intakes on the sides of the engine compartment. Particularly noticeable in this view is the circular hatch in the turret for the 'Kurbelmast "P", a telescoping winch mast bolted to the floor of the fighting compartment that extended through a hole in the roof of the stationary turret. The vision port for the second radio operator is open. Full details can be found in *Panzer Tracts* No. 3-4. **W.Auerbach**

Another view of the Jagdtiger from s.Pz.Jg.Abt.653 that appeared on page 90 of *Panzerwrecks X*, this time with the travel lock and track guard lowered. It was photographed by Sgt. Ken "The Vernal Cowboy" Workman of 3802 Quartermaster Truck Company.

K.Workman via A.Christensen

The title to this photo is "No more noise" and shows Sgt. Workman with the dead Jagdtiger. 'Vernal' means 'youthful,' and the Jagdtiger could not have been more than a few months old when this photo was taken, although it has aged considerably in 'tank years.' This view reveals that the vehicle was a Befehlswagen, as identified by the armoured 'pot' over the left rear hatch that protected the porcelain base of the 'Sternantenne'. The fact that daylight is visible inside the rear doors indicates that the superstructure roof has been blown off.

K.Workman via A.Christensen

'Pop' Eichers and Rod Grey at their after war occupation station in Gauting. Here they pose with the Jagdtiger in a field behind the Gasthof zur Post that was being used as a motor pool. We can see how US recovery crews moved the massive beast; by freeing it of its tracks and hitching tow cables to the drive sprockets. **R.Grey via A.Christensen**

A photo from the John Malan Heslop collection at Brigham Young University showing Mr. Helsop leaning against the dummy barrel of a training tank seen on page 80 of *Panzerwrecks X.* Thanks to him we finally get a glimpse of the front of this intriguing vehicle. In the background is a Panther Ausf.D chassis (note road wheels with 16 rim bolts). **Inset:** A similar training vehicle in happier times. It appears that the 'gun' was fixed in position. For all its size, the wood gas generator provided only the most feeble substitute for proper fuel.

1x BYU, 1x W.Markowski

A group of badly rusted Panthers. Some have the twin cooling pipes next to the left exhaust pipe, and at least one is without a muzzle brake. The headless Panther shown in the background is likely to be the same vehicle as pictured on page 81 of *Panzerwrecks X*.

3x L.Archer

These three photos were taken by Karl W. Speckman in Braunschweig. **Main photo:** Curious G.I.s man a Panther Ausf.A. This example has the armoured hub cap on its drive sprocket and rain guard over its binocular gun sight. **Left:** A round into the side knocked this Panther out. **Right:** The driver's hatch plate is upside down, revealing the mechanisms for the hatches.

3x D.Brown

More wreckage in Braunschweig. A rebuilt Pz.Kpfw III Tp Ausf.G or H upgraded with, among other things, additional armour plate along the lower rear hull (note large bolts along bottom edge). The rear armour on the engine compartment was not similarly upgraded, and an armour piercing round punched right through it. The turret has been fitted with 'Vorpanzer' and a stowage bin. Using parts from other Panzer may explain why the 'Balkenkreuz' on the rear plate does not line up. Note that the smoke grenade rack that normally appeared on the right rear plate is also absent.

L.Archer

Another rebuilt/upgraded Pz.Kpfw.III (5cm) Ausf.G or H, likely from the same unit because of the similar style of tactical number, although this one has not been tropicalized with raised air covers on the engine hatches. The turret has been upgraded with 'Vorpanzer.' Note the mix of early drive sprocket with late idler. **D.Brown**

This page and opposite: Front and rea of the same Pz.Kpfw.III Ausf.J Tp. Not the white corners on the mudguards. Th photo on the left gives the location a Braunschweig, the photo on the right a Erlangen.

1x L.Archer, 1x W.Auerbac

Pilsen airfield. A very unusual hybrid: a 7·5cm Pak 40/3 from a Marder III mounted in a 15cm sIG33/2 (Sf) auf GW 38(t) Grille superstructure (misidentified on page 48 of *Panzerwrecks X*.) Normal superstructures on Marder IIIs had a greater slope and different rivet detail, a cross bar over the gun and the antenna base on the right side of the vehicle. The Pz.Kpfw. Is may have served as munitions vehicles. The circular hole in the hull was also seen on the Pz.Kpfw.II on page 85 of *Panzerwrecks X*. **Opposite:** Ssgt. Robert D. French, Co. C, 7th A.I.B., 8th Armored Division, on a kleines Kettenkrad; 8·8cm Flak 37 on a mid-production Pz.Kpfw. IV chassis, and Jagdpanzer 38s.

1x D.Brown, 3x L.French

A clean and pressed PFC with the 8th Armored Division poses in front of a Panzerspähwagen AMR-35 (f) armed with a 7·5mm Hotchkiss MG. This vehicle is number '06' or '16,' and ties in with the vehicles on page 48 of *Panzerwrecks X*. Note the 'CS' painted on its nose, indicating that this light tank was used by Czech insurgents. The numbered bays in th background contain two softskin vehicles that we can see.

T.Mannin

A G.I. stands on the Grille/Marder hybrid to survey the airfield's assortment of AFVs. The mSPW (Sd.Kfz.251) carries a black stencil and has a very subdued 'Balkenkreuz' on its side and lacks a forward MG shield or swing arm for the rear AA MG. In the final months of the war it was permitted to paint the 'Balkenkreuz' in red. Behind it are two versions of the schwerer Panzerspähwagen (Sd.Kfz.234): a schwerer Panzerspähwagen with 2cm Schwebelafette (centre) and a schwerer Panzerspähwagen (5cm) (Sd.Kfz234/2).

L.Archer

This vehicle was covered by ETO Ord. Tech. Intell. Report No. 307 dated 8 June 1945 "German 8-wheel Armored Car Mounting 2cm KwK Gun." Observations by Capt. D.M.Gilles, Ord.Tech Intell. Hq.Co Z, ETOUSA. *A German 8-wheel armored car, Sd.Kfz.234. with a 2cm KwK 38 gun on a pedestal mount in the fighting compartment has been encountered. The chassis is that of* [a Sd.Kfz.234/3 with its armour cut down, as deduced by Mr. Archer based on internal stowage and rear base for AA MG swingarm]. *...the gun and gun mount are those used on the medium armored semi-track vehicle Sd.Kfz.251/17. No Sd.Kfz. number was marked on the vehicle examined. The fighting compartment is not fitted with a turret, the gun having 360 degree free traverse on the pedestal mount. The gunner is protected by the shield which traverses with the gun. ...A wood strip is placed over the top edge of the front plate to protect the gun while traveling. The mount provides free elevation from -5 to +60. The base of the mount is bolted to the frame of the vehicle. It is constructed of steel plates welded to form an "H" shaped structure with the side members parallel to and above the longitudinal members of the vehicle frame. The tubular upright on which the gun is mounted is welded to the center of the base.* In other words, it was a 2cm Schwebelafette on a cut down Sd.Kfz.234/3. **NARA**

Another portion of the vehicles assembled at the airfield, and a good idea of how many of them were brought there - under tow. Judging by its battered front end, this shop worn schwere Zugkraftwagen 18t (Sd.Kfz.9) may have pushed as well as it pulled. Our G.I. is probably taking a breather before he has to start the climb back down from his perch i the driver's seat. The connotation of the 'B' is unknown. The 7·5cm Pak 40/3/Grille hybri can be seen in the background. **D.Brow**

his photo, from a veteran of the US 83rd Infantry Division, clearly shows just how massive he schwere Zugkraftwagen 18t (Sd.Kfz.9) was and makes for a good comparison with the previous photo. The markings on the engine cover are Allied. The box on the side of the driver's seat was a battery warming box, critical for cold temperature operations. **AMC**

Our G.I. poses with a very colourful Marder II. Mud obscures the hull and suspension, but the road wheel on the track guard gives an indication of the colours beneath the dirt. The markings in white were added by the Allies, '431' may refer to the 431st Ordnance Motor Vehicle Assemble Company. This 7·5cm Pak 40 auf Fgst.Pz.Kpfw.II has the rare later styl cap for the recoil cylinder. Last issued to troops in July of 1943 but this example looks t be straight from the factory. It eventually wound up at the Patton Museum. **2x AM**

21/1/1
(2)

A beautifully photographed Sturmgeschütz III with free swinging, centrally pivoted 'Schürzen' plates scalloped along their bottom edges protecting the hull and running gear and rectangular plates protecting the superstructure. The 'Schürzen' may have been overlapping when they were spray painted to achieve the unique geometric shapes. Note the perforations in the skirts on the opposite side. An AP shot has skimmed the front of the mantlet, hit the top of it and wrecked the commander's cupola. This StuG III has a coaxial MG as well as a 'Rundumfeuer' MG over the loader. It was lost to artillery fire on the outskirts of Braunschweig (Brunswick). **2x AMC**

A Sturmgeschütz IV with intricate brackets for supporting its 'stepped' 'Schürzen' plates. The 'Schürzen' were able to slide outward on channels to accommodate wider winter tracks, and it is believed these plates were pivoted to swing as well. This StuG IV had extended hull side tow points and 'Rundumfeuer' MG but lacked sun shields on the driver periscopes and coaxial machine gun. Hints of a camouflage pattern show up on the track guard and elsewhere. **AMC**

A burnt out 10·5cm Sturmhaubitze 42 with standard side skirts that follow the contour of the superstructure closely. An internal explosion has shifted the roof and tossed the commander's cupola to the side. Fire has burnt the rubber road wheels and caused the torsion bars to fail, lowering the suspension. The original caption read: *Knocked out heavy tank, Germany*. The location is believed to be Bad Hosburg, Germany.

AMC

Hummels produced after February 1944 had the larger, full width driver's compartment although there appears to be no hatch where the G.I. is sitting in this photo. Bright sun reveals a high contract camouflage scheme. Markings consist of a tactical number '322' and a 'Balkenkreuz' on the side. **AMC**

An example of the penultimate Pz.Kpfw.IV Ausf.J with most of the very late modifications visible: extended hull side tow points; three steel return rollers; thickened armour cover over the turret ventilator, with a crescent shaped cut out to allow 360° firing of the 'Nahverteidigungswaffe' and a base for the 'Orterkompass' on top; bases ('Pilzen') on the turret for mounting a 2 ton jib boom and holders for the gas detection panels on the turret roof and gun mantlet. It is assumed the gunner's visor in the turret front has been deleted as well. Like most of the Panzers defending the Reich in 1945, this one is covered with pine boughs. **P.Johnson**

Not a view you're likely to see again anytime soon: The remnants of the SS-Pz.Rgt.1 were pushed from the road into a river bed in the vicinity of Waidhofen an der Ybbs and Weyer-Enns in Austria. At least two Pz.Kpfw IVs and one Panther are resting on their sides, jammed together so tightly that our G.I. with the carbine (Pfc Robert Burns B.Co., 42nd Tank Battalion, 11th Armored Division) is not worried about their tipping over. In total 2 Pz.Kpfw.IV, 3 Panthers, 2 trucks and 2 carts ended their service careers here.

K.Shea

PFC David 'Bruce' Lovett of HQ Co., 66 Infantry Battalion, 12th Armored Division poses with a Pz.Kpfw.IV Ausf.J fitted with the style of 'Schürzen' fitted prior to September 1944. The tank has extra protection courtesy of track links from different types of Pz.Kpfw.IV and T34 tracks. The horse shoe on the driver's front plate did not bring this tank much luck. Germany 1945.

B.Lovett

54cm Karlgerät Nr. V in First US Army area shortly after being sent to the front. This vehicle, once released from its railway transport, could travel short distances under its own power. This version had a twelve tooth sprocket and eleven steel road wheels. Its 60cm Mortar was exchanged for a 54cm barrel at Jüterbog in Nov 1944. **Opposite:** Railway truck and supporting truss. All photos from ETO Ord.Tech. Intell. Report No. 328. **2x NARA**

Deutsche Reichsbahn
Berlin 918 542 P
Bahnhof
Wustermark Vbf
Dessart
Berlin W 35
Zw Br Unt Bln
27.4.44

Front of 60cm Karlgerät Nr. II with seventeen tooth sprocket and eight road wheels. Note air cleaners immediately above opening in front plate and method of attaching railway supporting truss. Loading jack on railway truck is at right. The engine access hatch is missing. **NARA**

nterior of twelve tooth drive sprocket of Karlgerät Nr. V showing its planetary ;ear. The Maybach MB507 diesel engine that powered this massive vehicle vas rebuilt when Karl Batterie 428 returned to Jüterbog in Nov 1944. Ninety our track links made up each track. **NARA**

Suspension of Karlgerät Nr. V. A mechanism in the rear of the vehicle's hull could be used to raise and lower the hull directly onto the ground for firing. Karlgerät Nr. V had a driving range of 60 km with a full 1200 litres of fuel in its tank, but was so huge and slow moving, and needed such an extensive entourage of vehicles to support it, that it was easily preyed upon by Allied planes and artillery. Be that as it may, it was paired with other Karlgerä and sent both East and West to deal with high value targets that could not move at al

Opposite: Rear idler of vehicle Nr. V. Components such as idlers might be rebuilt or take from damaged vehicles. Note turnbuckles supporting the track. **2x NAR**

Rear idler and suspension of Karlgerät Nr. II. Note the track is still suspended for rail travel, and this vehicle had done a bit of travelling: It was overhauled with a new engine at Jüterbog in September 1944, assigned to Karl Batterie 428 in November, and shipped to Ob.West in December where it was damaged. It was returned to Jüterbog in January where it was repaired with parts from Karlgerät Nr. I, and shipped out again in March (Karlgerät Nr. I was the only one to share the same suspension with Nr. II.) **Opposite:** Fron sprocket and suspension of Nr. II. Based on its success at destroying Warsaw, a Karlgerä was ordered to Paris in August of 1944 but was diverted to Warsaw instead. **2x NAR**

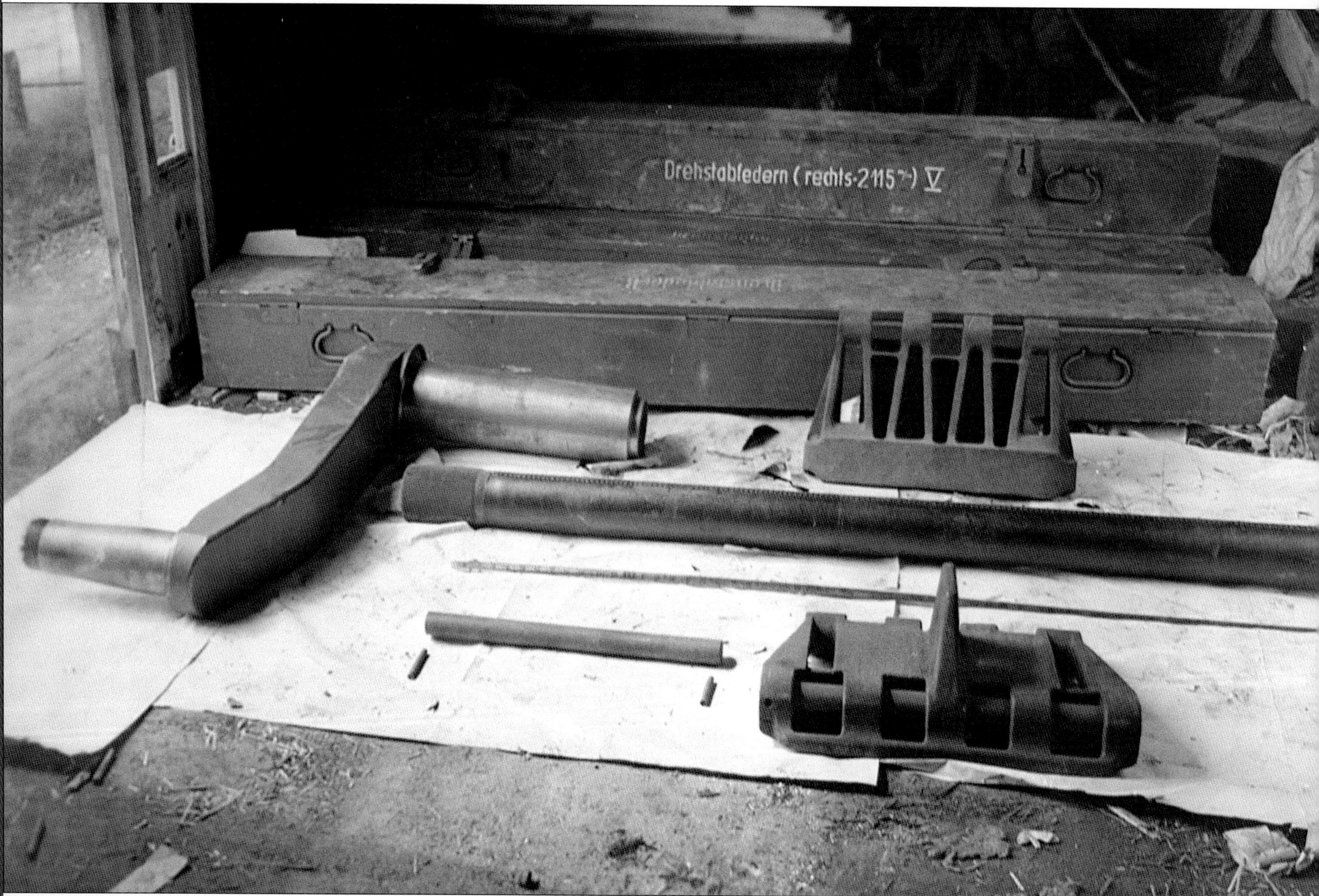

Crank arm, 2115mm torsion bar and track sections of Karlgerät Nr. V. Note boxes of spare torsion bars in the background. Mountains of spares had to accompany these vehicles when they were sent to an active front to ensure their serviceability. **Opposite: Top Left:** Top carriage recoil system and slides. The recoil system was robust enough to withstand the forces sustained firing several hundred rounds weighing about 2 metric tons apiece. **Top right:** Rear carriage truck with traversing mechanism and bearing. **3x NARA**

Bottom left: Interior of hull showing top carriage recoil system yoke locked in recoil position for travelling. **Bottom right:** What's left of the power train in the front compartment. Engine is at upper left and torque converter is at upper right.

2x NARA

Driver's compartment and controls. Early deployments of the Karlgerät in 1941 required a large emplacement to be dug out, levelled and camouflaged prior to the vehicle backing into firing position. Since traverse was severely restricted (4° maximum), the sunken position had to be wide enough so that the entire vehicle could manoeuvre within it. **NARA**

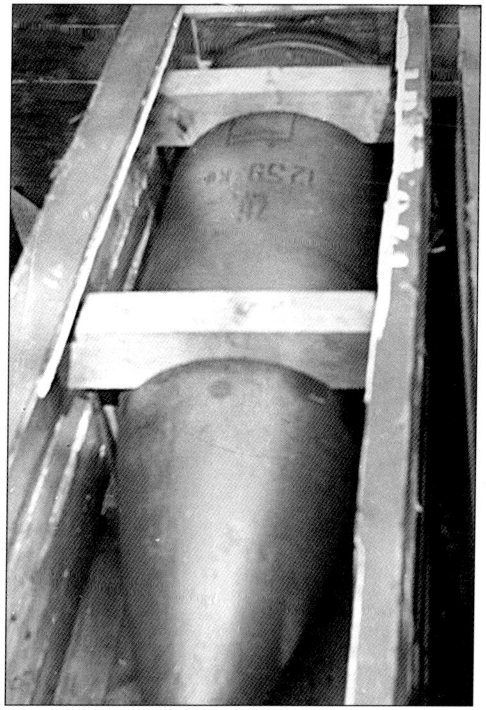

Top left: 54cm Karlgerät with breech open. The loading apparatus is in the foreground, and had to be removed before firing. **Top right:** 54cm shot tongs. 'Kettenglied' translates to 'track link.' **Bottom left:** 54cm projectile in packing crate. Concrete busting 'Betongranaten' rounds became available for testing in October but what targets were envisioned for it are unknown. High explosive 'Sprenggranaten' rounds weren't available for testing until November of 1944. **Bottom right:** 54cm cartridge case.

4x NARA

Elevating mechanism and bottom gun recoil system of the 54cm Karlgerät.

NARA

oading mechanism. Large subassemblies such as this had to be transported separately, aboriously hoisted into place, and then removed each time the weapon was to be fired. 'he 54cm and 60cm weapons required custom loading mechanisms. Firing was roughly one round every ten minutes, provided rounds did not jam or explode in the barrel. Between 15 and 18 'Kanoniere' were needed to service the weapon.

NARA

pposite: Sixteen wheel road transporter ('Culemeyer-Straßenfahrzeuge'). At least three f these were needed to transport the upper assemblies: gun barrel, recoil mechanisms nd loading tray. Each trailer in turn required a prime mover. **This page:** A captured photo f a Munitionsschlepper based on a Pz.Kpfw.IV Ausf.E with loading crane in travelling position. At the extreme right is the 'Greifer' (shell gripper) similar to the one seen on the top right of page 91. New grippers had to be developed each time a new round of ammunition was developed, so that by 1944 four different types in all had been created to deal with the light and heavy shells in each calibre. **2x NARA**

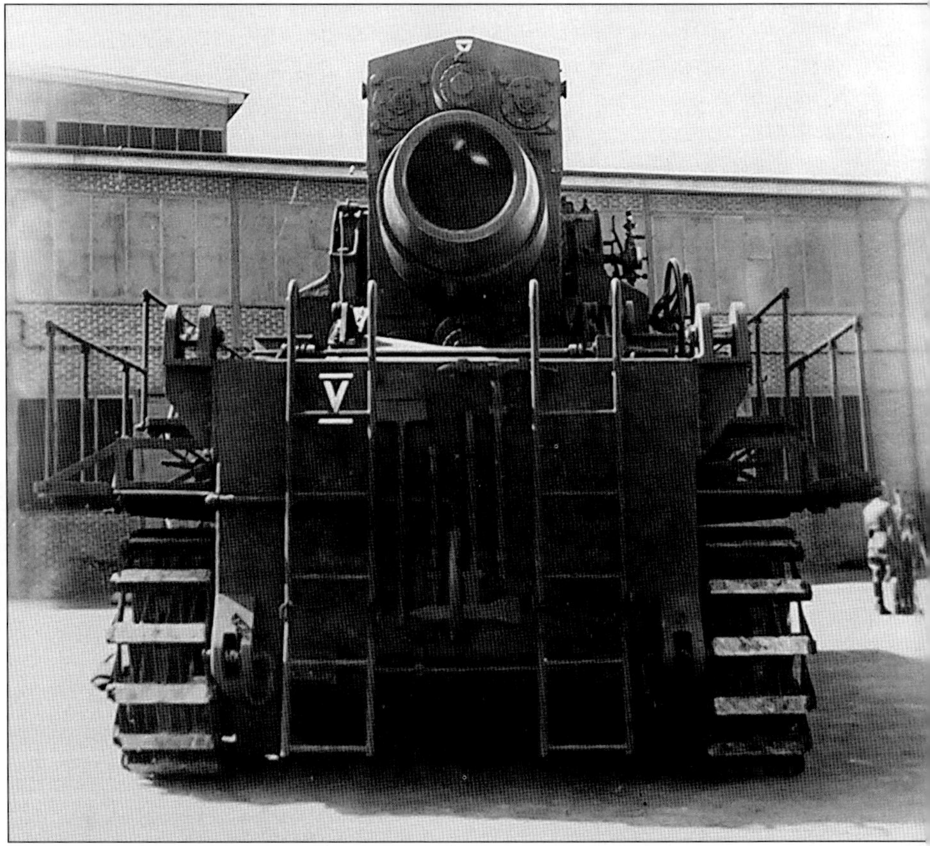

Top left: Carriage with dual recoil system on Culemeyer-Strassenfahrzeuge. **Top right:** Fully assembled 60cm Karlgerät ready to be driven to its firing position. The muffler and exhaust pipe ran the length of the vehicle. **Bottom left:** Breech, elevating and traversing handwheels and sight assembly for 60cm Karlgerät. **Bottom right:** Rear view. **4x NARA**